UFOs + DNA = LIFE

Arguments in support of Rael testimony
from his encounter with the extraterrestrials
who created all life on Earth

Marc Letourneau

First Edition

ISBN 978-0-9829451-4-8

Cover

Concept design: Marc Letourneau

Realization/main objects and vectorization: Sylvain Bernier

Realization/beautification: Claude Chevey

(Many chapters of this book were part of a published book printed in March 2010 and officially launched on 24 April 2010. This previous book was retired in 2017 because it was thought that some information was perhaps not appropriate for the public at the time.

Table of Contents

Acknowledgements

Though NASA is hiding essential information from the public, I am grateful to this organization for the pictures from space it has shared. Thank you to my friend Larry who reviewed some addition to the original text. My warm thanks also go to Rael, whose message helped me tremendously to understand the origin of life, the universe as well as myself.

Apology

I apologize for the medium quality of the writing as English is not my native language (it is French). But I hope the text is clear enough so the content can be understood.

Preface

I was raised in a French Canadian Catholic tradition in Quebec province in Canada. At 12 years old, I changed my view and turned to biology and physics to answer some fundamental questions (Who are we? What is the origin of life? Is the Universe finite?). Then at the age of 20 I studied the UFO phenomenon. At 21, I examined the message given to Rael by extraterrestrials. This reading changed my previous vision of the world.

In this book, I will try to explain how I structured the various pro and con arguments relative to this message from space as well as the current paradigms in science and philosophy. But, thanks to this message from space, here briefly is how I finally answered the fundamental questions:

Where are we? We are in an infinite universe.

Who are we? We are conscious entities that are capable of sensing the universe from the inside out; capable of being aware of our own entity. We are composed of atoms and all matter and energies that are part of the universe, but organized in such a way as to make ourselves, sentient beings.

Where do we come from? We come from other conscious entities propagating life in the Universe. The planet Earth has been terraformed.

This text may help others to understand these answers, and in turn, help them make their way through life. Or at least it can help people understand the Raelian Movement and why there are people supporting this Movement.

I – Tabula Rasa

On the tiny spaceship that is Earth, humans have never ceased to wonder about the medium in which they are immersed and of which they are a part. To the question "where are we?" humans on all continents have formulated many answers, thereby accumulating knowledge over millennia.

Can we, therefore, conclude that we know more now than we used to? Of course we can! Do we not cross the seas in a matter of hours? Have we not sent satellites into space, some of which have gone beyond the limits of our solar system? Indeed, we have sufficient understanding of our environment to be able to achieve such things. However, as astonishing as it may seem, the understanding that today's scientists have about the Universe remains superficial. They may have enough accurate knowledge about the effects of gravitational force to deduce mathematical formulas and these can be used to calculate everything (or almost everything) related to trajectories. But does this mean they know what the mechanism of this fundamental force is? Actually, we contend that they know absolutely nothing about it! What we have are theories that have more to do with the imaginary than what would be the still unknown reality.

One has to realize that Newton, the "discoverer" of gravity, did not understand how it worked. In relation to Newton, Professor Dr. Pavel Kroupa (Bonn University) wrote:

> Although his theory does, in fact, describe the everyday effects of gravity on Earth, things we can see and measure, it is conceivable that we have completely

> failed to comprehend the actual physics underlying the force of gravity.[1]

A theory would most probably increase its chances of being true if it were based on true postulates. For instance, beginning with the assumption that absolute emptiness exists can lead to theories that would be completely different from those we may construct if we assume that absolute emptiness does not exist because interpreting the facts will vary depending on the assumption we make.

One clue that suggests that a theory is not based upon good postulates is to be found in cases where there is essentially no explanation for a phenomenon; calling a phenomenon a "force" without explaining its physical mechanism is equivalent to saying that the phenomenon is supernatural or immaterial. It is exactly what we observe in our current theories regarding the structure of the Universe and its "origin."

None of the forces in the Universe have been adequately explained and they need to be looked at with a fresh and new vision. It can be stated that the postulates on which we rely for our understanding of the Universe are, at the very least, partially false since they point to quasi immaterial forces. Therefore, these forces become weird, almost magical or supernatural, rather than scientific ones. Indeed, if no scientist knows the materialistic mechanic of how gravity or magnetism works, we are in the realm of magic. This does not represent real understanding. It is mysticism

---

[1] Metz, Manuel; Kroupa, Pavel; Jerjen, Helmut: Discs of Satellites: the new dwarf spheroidals (Monthly Notices of the Royal Astronomical Society 2009; doi: 10.1111/j.1365-2966.2009.14489.x), cited in physorg.com (http://www.physorg.com/news160726282.html).

disguised as science. Although it served to introduce the subject of this book, we will leave the question of the fundamental forces for another time.

We believe that to better understand the essential of our universe, it is necessary and relevant to start over from the beginning from a first principle process and build on it. Taking a tabula rasa[2] approach that is mainly philosophical will allow us to clarify things from the "ground up." Starting from scratch, we will perhaps be able to find a way to answer some questions.

Three basic postulates

In addition to the basic assumptions of science, which are that the human mind is able to understand the Universe[3] and that the supernatural does not exist,[4] let us state the following postulates:

First postulate:

> The Universe is to be defined as: everything that exists, known or unknown, visible or invisible.[5]

Second postulate:

> What exists "occupies" a volume, however small.

---

[2] Blank slate.

[3] This simply assumes that the mind is flexible enough to understand anything, especially when multiplied by the diversity of minds in humanity.

[4] A supernatural phenomenon is only an unexplained natural one.

[5] Visibility is relative to what we can perceive with our eyes or with the help of scientific instrumentations.

Indeed, how could something that does not occupy any volume exist? If one believes such an idea to be true then it may be one has not thought sufficiently about the mathematical limits that describe any phenomena. For example, the mathematical concept ascribed to the number zero does not necessarily mean that it is represented in space volume. It is a *relative* concept when applied to the real world. For instance, having zero apples in a basket does not mean there is absolutely nothing in the basket. There is normally air, for one.[6]

Third postulate:

An absolute vacuum cannot exist.

If an absolute vacuum were to exist, even in between the dispersed atoms still found in deep space[7] or in between any of the smaller known particles in space, it would create an unacceptable paradox because an absolute vacuum, as defined as the absence of anything, literally means

---

[6] If we cannot imagine the mechanism of a physical phenomenon, we simply do not understand it. A good way to check if we can imagine it is to try to make a rough drawing of it, which would allow us to check a part of its validity. This does not mean that being able to draw something proves its existence, but it is a minimum requirement.

[7] Some may connect this idea with "free space," which is a concept that means empty of matter (referring to known elements), that is, according to standard physics, unattainable because there is always some kind of elementary matter, whether an atom, a photon, etc., in a volume of space. But the question we are discussing in this chapter is more whether or not: Is there a physical background fabric of space with no empty space or is it absolute emptiness containing these elementary particles.

"nonexistent." Thus, to say that "nonexistent" exists is blatantly illogical.

Here, we are rediscovering an antiquated idea that fascinated Greek philosopher Parmenides. Whereas the notion of a vacuum caused interminable discussions, Parmenides proposed a seemingly unassailable argument, which became popular in his time:

> To be is; not to be is not.

Of course, there is nothing better than a tautology to point out the obvious – or the meaning of words, for that matter. What this means is that "nothing," or absolute emptiness, cannot exist!

Unfortunately, this paramount tautology, which is impossible to circumvent, was abandoned when Rutherford's atomic model came forth. Rutherford's model suggested that the atom was mainly a vacuum. This concept of vacuum was a total contradiction to Parmenides' notion of the Universe.

This "vacuum" gradually came to be regarded as absolute. In comparison to Rutherford's hypothesis, Parmenides' argument began, from that point on, to be considered a mere curiosity. However, we are suggesting that to abandon Parmenides' ideas was a big mistake.

Let us recover and restate this principle because it is an essential one. "Absolute vacuum" means there is nothing in any given place. But "nothing" means "no thing" or "nonexistent!" The paradox is that we cannot say that "nonexistence" exists or that "no thing" is a "thing."

and

NOTHING = NONEXISTENCE

therefore,

ABSOLUTE VACUUM = NONEXISTENCE

"To be is; not to be is not."

Conclusion: An absolute vacuum does not exist.

This indicates, hopefully clearly to enough of the readers, that we cannot state the existence of an ABSOLUTE VACUUM without contradicting the intrinsic meaning of the words. Words convey meanings and we cannot deny them on the pretext that they are mere semantic tricks. Semantics are indeed the very basis of any precise language and the foundation that leads to the formulation of assumptions and to the verification of theories. To ignore this only increases the existing confusion and allows discourses that do not make sense.

That an absolute vacuum prevents the existence of things has been all but forgotten and obscured by wordplay. By replacing "nonexistent with the expressions "vacuum" or "void," we have transformed the negation of existence into the affirmation of an underlying "something," which gives us the illusion of existence. It is a semantic pitfall, which, if ignored, can lead us into error and take us down the wrong path. This is what happened to our ancestors and, unfortunately, this same semantic pitfall has persisted until now through education. Building the foundation of the house of physics on quicksand and perpetuating this legacy through education has led us to the current state of affairs in which our theories about the Universe may have sunk into falseness, partially at least.

A very important consequence of this new understanding, if we want to follow its logic, is that the Universe is not an empty space containing objects that fill a void. On the contrary, the existence of things "creates" space. One may want to reflect on this for a while…

Now that we have established a rock-solid framework, we can start building a bit of foundation. We can now answer the question that has appeared unsolvable to many: "Why are there things rather than nothing at all?" This very interesting question can be poetically expressed by another famous question: *To be or not to be*?[8]

Indeed, philosophers and scientists throughout the ages, although they were not questioning the existence of absolute emptiness itself, have wondered, in a general sense,[9] "Why is there something rather than nothing?"

What happened is that wrong postulates have led the most brilliant scientists to dead ends as Stephen Hawking did:

> The universe can and will create itself from nothing… Spontaneous creation is the reason why there is

---

[8] Hamlet (Shakespeare).

[9] It is a fact that the Universe is filled with things (planets, molecules, etc.) we know exist and the question posed by some scientists was whether the Universe could once have been filled with nothing at all.

> something rather than nothing, why the universe exists, why we exist...[10]

The problem with this conclusion is that when you have nothing, you will have nothing forever... Stating that something can come out of nothing indicates a lack of understanding of the word "nothing." It is utterly illogical and similar to magic, but disguised with a science-coated wording.

However, the answer, as we have shown, is simply that "nothing" cannot exist. Therefore, what exists does not have the choice of existing *forever*, just as "nothing" does not have the choice not to exist forever! To be is; not to be is not! That is all there is to it!

We are thus forced to realize how extraordinary this new vision of the Universe is in term of answering fundamental questions. The Universe is required to exist without any void, and since everything should be linked to everything else, it makes sense for all the parts to be working intrinsically together in a logical manner, like cogs in a watch.[11]

---

[10] Hawking, Stephen, cited in cnn.com, 3 September 2010, from book: *The Grand Design*:

http://www.cnn.com/2010/WORLD/europe/09/03/hawking.god.universe.criticisms/index.html?hpt=T2#fbid=XONZ2S30moy&wom=false

[11] Joon Hyeop Lee, Mina Pak, Hyunmi Song, Hye-Ran Lee, Suk Kim, Hyunjin Jeong, "*Mysterious Coherence in Several-Megaparsec Scales Between Galaxy Rotation and Neighbor Motion*," arXiv:1908.10972v1 [astro-ph.GA], 28 August 2019. https://arxiv.org/pdf/1908.10972.pdf

I - Tabula Rasa

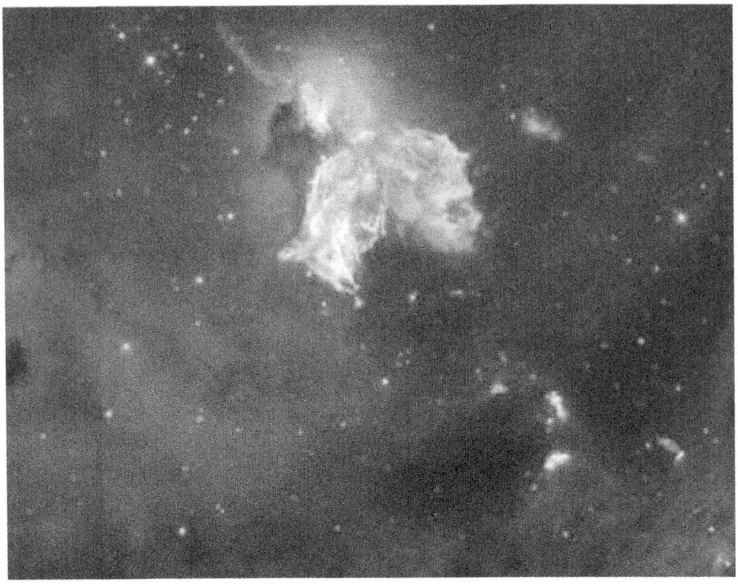

II – Infinity

> More than any other question, that of infinity has always tormented the sensitivity of men; more than any idea, that of infinity stimulated and fertilized their reason; but more than any other concept, that of infinity requires to be elucidated… David Hilbert

The immediate question that arises, with respect to the above-mentioned three postulates, is: "What type of Universe do we live in?" One can understand that there are only two possibilities that describe the fundamental nature of the Universe – either it is finite or it is infinite. Therefore the essential question is: "Is the universe finite and bounded, or is it infinite and boundless?"

This marvelous question, which contains two mutually exclusive exact opposites, facilitates the answer because when we are faced with only two exclusive alternatives and no others, it is sufficient to eliminate one to obtain the solution. The answer must reside in the one that remains. Since the Universe cannot obviously be both finite and infinite simultaneously, demonstrating that one of the two alternatives is absurd should be enough to solve the enigma.

With respect to the first alternative – that the Universe is finite – we can proceed by stating that if the finite universe were to exist, it would be limited in space and in time as follows:

1) It would have a limited circumference.

2) It would consist of elementary particles.

3) It would also consist of elementary sections of time.

4) It would, by definition, have a beginning and an end, the greatest length of time of all.[12]

A finite universe would have an outside limit beyond which there "would exist" an absolute void, or absolute vacuum. This already leads us to a paradox. Since there is no absolute void,[13] there cannot be an outside limit to the Universe.[14]

Furthermore, the space between atoms, according to this finite universe, would also be made up of an absolute vacuum! Yet, again, an absolute vacuum cannot exist. The structure of the finite universe imposes an absolute vacuum and this is precisely what destroys the argument. A finite universe in space is thus impossible within the framework we have defined.

Using a parallel argument, we cannot envision how a particle could not be composed of smaller particles. If you break a particle apart you inevitably obtain smaller particles that make up the very fabric of the said particle. But, if the smallest particle did exist, it would have to have a volume equal to zero; otherwise it would theoretically be divisible. Nevertheless, a voluminous universe could not be composed of elementary particles with no volume. Zero +

---

[12] Does it make sense for a universe to be finite toward the large and not toward the small or for a finite universe in space to be infinite in time or vice versa? This does not fit at all with the accepted concept that the absolute void cannot exist, as will be shown.

[13] Postulate III.

[14] Liza Zyga, Phys.org, "*Model describes universe with no big bang, no beginning, and no end.*" 29 July 2010. https://m.phys.org/news/2010-07-universe-big.html

zero equals zero, not two! Moreover, a volume of zero contradicts the second postulate described previously. It is not a coincidence that any description of an elementary particle that would be indivisible is intrinsically mystical and is doomed right at the start logically. Elementary particles do not make sense. They have to be composed of smaller particles.

From this simple demonstration, we can conclude without hesitation that the Universe is infinite in space, i.e. the particles are made up of particles, which, in turn, are made up of particles, etc., in a fractal regression, toward the infinitely small. Conversely, our solar system is a very small part of our galaxy, which itself is a very small part of clusters of galaxies, themselves a very small part of something else, in a fractal progression toward "the large." This is a simple extrapolation of what we observe in the visible part of the Universe. All things are made up of something smaller and are part of larger things. It is a generalization of the observable.

In between every particle, from the infinitely small to the infinitely large, what would seem to be empty would in fact be filled with "something," making up the fabric of the Universe. Something is invisible because it exists at an infinite continuum of space levels, which at some point is so small that it is imperceptible to our senses and to our most sophisticated and current scientific instruments, such as microscopes and telescopes. However, since perfecting science and making new discoveries progresses on an infinite continuum, the objects that are currently imperceptible will gradually become "visible," thereby making the invisible visible or detectable on an infinite continuum of relative invisibility.[15]

The temporal existence of the infinite Universe is also infinite, i.e. it did not have a beginning, nor will it have an end. Indeed, having a beginning or an end would mean that before this beginning or after this end nothingness or the absence of anything at all "would have existed or will exist." This simply translates to more nonsense! Nothingness only exists as a word, but the object it defines excludes itself from the real world by definition: "Nothingness" cannot exist: it is "nothing"!

Therefore, the Universe has always existed and will continue to exist *ad perpetuum.* Furthermore, using the current concept of time for the practicality of this demonstration, "time" within the continuum of smallness will, in an infinite universe, divide itself *ad infinitum.*

A smallest element of time cannot exist because a smallest time would still be divisible; therefore, it cannot be the smallest time. To be clear, a bona fide time cannot be formed from an elementary time equal to zero, which would be the only option for the smallest time to really "exist." Since anything greater than zero is divisible, there cannot be a smallest period of time.

Because this is beyond the scope of this book, no further discussion is required at this time. Nevertheless, it is worth stating that an infinite continuum of space toward the infinitely small implies there must be an infinite continuum of "time" toward the infinitely small.

---

[15] Infinity in space is a very strange concept, and we will demonstrate shortly how it can be proven experimentally.

This concept of infinity in space is not new. It was mentioned in a number of ancient writings. The concept of infinity in space was revived by Carl Sagan in his popular book, *Cosmos*.

> There is an idea – strange, haunting, evocative – one of the most exquisite conjectures in science or religion. It is entirely undemonstrated; it may never be proved. But it stirs the blood. There is, we are told, an infinite hierarchy of universes, so that an elementary particle, such as an electron, in our universe would, if penetrated, reveal itself to be an entire closed universe. Within it, organized into the local equivalent of galaxies and smaller structures, are an immense number of other, much tinier elementary particles, which are themselves universe at the next level, and so forever – an infinite downward regression, universes within universes, endlessly. And upward as well. Our familiar universe of galaxies and stars, planets and people, would be a single elementary particle in the next universe up, the first step of another infinite regress.[16]

This is right and so beautiful!

We may not yet have experimental evidence of the Universe being infinite in space and in time, but between two types of unproven universes, we should choose the one that makes the most sense. This is a completely acceptable process.

We suspect that an experimental proof that the Universe is infinite in space is possible, and that this proof will involve recognition of the repeated structure of particles of all sizes. If galaxies or clusters of galaxies do in fact represent atoms

---

[16] SAGAN, C., Cosmos, Montréal, Sélect, 1981, p. 265.

of a much more gigantic world, it will be possible one day to observe that the configuration of these atoms does form recognizable molecules as such. The same reasoning can be applied for small particles that compose our universe. If atoms are in fact small galaxies or clusters of galaxies, it will be possible one day to observe the characteristic behavior of stars and planets involved in the formation of these particles. Until then, due to a lack of empirical evidence, it is better to trust what makes sense. Because this depends on the brains of those who contemplate such possibilities, as always, it will be a matter of opinion.

If some people purport to have scientific evidence of the Universe being finite, we may later discover these proofs to be mere illusions. Sadly, many people do not differentiate between the interpretation of a fact and the fact itself. For this reason, many so-called "proofs" will turn out to be mere interpretations of certain facts. While a fact can often be interpreted in different ways, one interpretation could have been chosen simply because it supported a finite universe, the underlying philosophical postulate of many theories. Otherwise stated, scientists who are subconsciously looking for a beginning or a starting point will have a tendency to interpret facts based on this premise. Of course, looking for a beginning in anything suggests that the model that will be chosen will imply finiteness.

Another reason that the concept of an infinite universe does not attract a great number of people, including scientists, is the effect on the psyche of the limitation of human sensory and cognitive capabilities. We do not hear or see (not to mention the other senses) beyond a certain distance, and therefore, the existence of these sensory limitations influences the way we see the world. Moreover, infinity in time and space is something impossible to imagine. Indeed, the brain is limited in its capacity to imagine space and distance: It cannot, with its limited number of cells, imagine what an infinite distance would look like. The consequence of this is that scientists usually try to stay away from what they cannot imagine.

Adding to the difficulty is the mathematical formulation of infinity in theoretical models of the Universe. Because it is difficult to manipulate, physicists would like to discard the concept, but infinity, being as rebellious a concept as it is, always bounces right back.

> Physicists try to eliminate infinity, which emerges from their theories: infinitely large time and space, infinitely small matter, singularities of black holes and the Big Bang... but the elimination of infinities leads to more complete theories, which in turn generate more infinities. Infinity is a sphinx that questions physics and shows new ways forward.[17]

Many philosophers and scientists have written that an infinite body is unconceivable. But perhaps it is necessary to differentiate between imagination and logical understanding. There might be a way of partly understanding the concept of infinity through imagination, by understanding the repetitive nature of structures. If the same structures of the Universe are found in the infinitely small and in the infinitely large, the loop is connected and the need to understand can be satisfied by imagining a complete loop while rationalizing and generalizing the rest.

---

[17] LUMINET, J.-P., Lachièze-Rey, op. cit., p. 62 (translated from French by the author).

If we find the equivalent of solar systems in subatomic particles, then we are clearly looking at an infinite continuum, especially if we see traces of the same intelligent manipulations on those particles that we will eventually be doing on our star systems in the future. Indeed, we might be able to move solar systems in the future. Hence, it is through the repetitive nature of structures that we will be able to prove infinity through experimentations.

The concept of infinity is indeed puzzling. Imagine for a moment that you want to teach the mathematical concept of infinity to children who have never heard about it. Their initial reaction is colored with incredulity. After all, it is a very strange idea. Perhaps you recall your own reaction to it when you were young. The same kind of reaction is likely to occur with any person who is told for the first time that the Universe is without a beginning or an end.

However, the impression that the Universe is finite is a persistent one, especially because our education has been saturated with either monotheistic religion or the Big Bang concept. Indeed, teaching that God created the Universe can only hinder students' ability to develop critical thinking and restrict intellectual growth. A created universe always needs a beginning,[18] but many people do not realize this obvious logic. Furthermore, even if a person rejects his or her religion, what often remains is the perception of a finite universe. Even if one replaces one's religious belief with an adherence to a scientific theory, this theory will probably have the same underlying notion of a finite universe.

[18] More is written on this subject in this chapter.

Perhaps some scientists unconsciously seek to maintain this idea of a "beginning" when discussing or defending such theories as the Big Bang. The difficulty the human mind has with the perception of infinity seems to come from the psychological reaction that it triggers, which prevents a logical analysis of the problem.

Once we can agree that infinity is the basic structure of the Universe, the logical consequences we can draw are quite fascinating. Not only does this agreement make it possible to exclude other models that imply a finite universe, but it may also enable us to explain the mystery of the forces of the universe, which is not the purpose of this book.

One major logical consequence of understanding infinity is that there can be no God that created the Universe. The answer to the question of God's involvement with the Universe is as simple as this: An infinite universe cannot be created, thus there is no God, period! That's one big question that is now resolved (for those who agree with the reasoning anyway)!

Again, an infinite universe has neither a beginning nor an end. If there were a beginning to the Universe, defined as "everything that exists," there would have to have been nothing, absolutely nothing, prior to that beginning.[19] We now understand that "absolutely nothing" cannot exist, but, for the sake of our argument, even if it could exist, we would end up with "nothing" *ad infinitum* because when there is nothing, nothing can ever occur, and nothing would "exist" endlessly; there would not be a universe. But again, the concept of nothing is not realistic and would exclude

[19] This rules out the preexistence of God.

any eventual creation of "what exists." Infinity of what exists is the only logical answer.

The Big Bang theory can also be checked on its basic validity. What was before this Big Bang? The theory holds that the whole Universe was "compacted" into a small particle. But where did that small particle come from? Did it have a beginning? This implies that there was nothing around that particle. The Big Bang theory does not answer this question.

That is the failure of the very scientists who want to limit themselves to traditional ideas of what being a scientist is. They could venture to think outside the box and embrace a new perspective and be prepared to consider basic philosophically sound ideas upon which theories and experiments could be built. Therefore, it is likely that in addition to resolving questions relative to the Big Bang theory, they would have resolved the question of the origins of the Universe a long time ago.

If the original particle that exploded in the Big Bang existed, it would be surrounded by an absolute void, since the Universe would be inside the particle and not outside of it. However, this means this particle would in fact be surrounded by emptiness or a void, and we are plunging back to the same paradox, which we have already established to be irrational.

Of course, some proponents will say that a "Big Bang" could have happened in one part of an infinite universe side by side with other universes experiencing their own Big Bang. But what was between them? A void? That cannot be. But if it were not a void, then we are just talking about explosions that occur in the Universe and such explosions do exist, as in supernovas. But that is not what the Big Bang theory is all about.

> There is a growing body of data and theory that questions whether the Universe may have begun with a Big Bang 13.75 billion years ago. Several leading cosmologists, such as Sean Carroll of CalTech and Neil Turok of Cambridge University challenge the prevailing model of a "Big Bang" and believe that in the future we will only look back in wonder at how anyone could have believed in a creation event which was refuted by so much evidence.
>
> The origin of the Big Bang, that is, the state of "existence" which resulted in a Big Bang, is a mathematically obscure state - a "singularity" of zero volume that contained infinite density and infinite energy. Why this singularity existed, how it originated, and why it exploded, has led many scientists to question and challenge the very foundations of the Big Bang theory.[20]

The stars emit a range of visible and invisible radiations, and since a redshift has been observed for most of them, scientists have adopted the position that these stars are going away from us. But I rather think that it is just that the higher frequencies are more easily blocked by dust and particles between Earth and the stars than because they are going away from us. That is a known phenomenon. Radio and sonar of lower frequencies project further away than those of higher frequencies. But this interpretation was not

---

[20] Kazan, Casey, *Does the Large-Scale Structure of the Universe Nix Big Bang Theory?*, 2010 www.dailygalaxie.com,

http://www.dailygalaxy.com/my_weblog/2010/05/the-big-bang-theory-fact-or-fiction-many-experts-say-fiction.html interesting as it did not lead to the desired universe: a finite one. So it was not "liked" and adopted by the majority.

Philosophically, we can conclude that there is neither a God nor a Big Bang, but rather Infinity! The belief in both God and the Big Bang is based on misinterpretations: the misinterpretation of the Scriptures (such as the Bible or other ancient texts), and the misinterpretation of certain scientific facts such as the radiation of stars shifting toward the red, in the case of the Big Bang theory. Ironically, the concepts of immateriality and absolute void share communality between theology and certain scientific theories in the sense that both concepts are unreal, supernatural, and are based on the idea that the immaterial (God or void) "things" exist.[21] These concepts are simply not defined physically; they are not really explained and this makes these concepts relegated, to some extent or other, to the realm of mysticism. The philosophical basis we have just established suggests that we eliminate the theoretical venues that lead to immateriality and the concept of a void in their absolute sense because both have no rational basis. Why build a house on quicksand when you can build it on solid rock?

In summary, the logical idea that there is no absolute vacuum alone implies infinity in space and time, because all the "space" in between planets, galaxies and particles would not be empty and therefore should "contain" something that is in turn composed of something else. If we

---

[21] Perhaps believing in the concept of immateriality or absolute void comes from a simple psychological reflex, which stems from an inability to visualize any of the operating principles underlying a phenomenon. It is a description of what we see and imagine: nothing.

do not see that "something," it is because it is beyond the limits of our perception.

What is also implied is that every element of infinity should be "in touch" with everything else,[22] on all levels, moving and vibrating in unison in a perpetual dance regulated by the laws of motion, which we try hard to understand while observing the visible effects. But in an infinite universe, there is only a small, limited part that we can and will be able to experience via our senses and their technological extensions (telescopes, microscopes, etc.). In every era of our scientific progress, there will always be a part of the Universe that will be too small or too large to be perceived by our instruments.

The intangible part of infinity is a kind of *relative* immateriality that has been, and can be, mistaken for absolute immateriality or an absolute void. Relative immateriality refers to matter we cannot experience with our senses or our technology. It does exist, however, in the same way that electromagnetic waves of the spectrum exist but cannot be perceived.[23] These electromagnetic waves can nevertheless be measured. Hundreds of years ago, we could not measure these waves but could only assume that they existed. The same applies to the smallest particles of matter and their associated waves that we cannot yet perceive. Although we will not be able to see the entire Universe (because of its infinite nature), we will continue

[22] Indeed, since there is no absolute void, what does exist has to be in touch with its surroundings.

[23] Pigeons and some other animals are thought to have a sensation of the magnetic field so they can somehow find their way around. Humans may also be equipped with such faculties but they must be dormant.

to see more and more of the matter in the infinitely small (and infinitely large) as our technology progresses. In brief, what is inaccessible in the infinitely small is not immaterial, but represents matter so small that we cannot experience it through our senses and technological equipment.

The infinity concept also destroys the concept of the speed of light being a limit. We will be able one day to measure micro vibrations that are much faster than light since these will be vibrations of much smaller particles, on a much smaller scale that are unperceivable to our instruments today. Infinity of space imposes an infinite continuum of speed.

Is the Universe conscious?

Some believe that the Universe itself is God. But why call it God? Why replace an established name with one that is more ambiguous and that contains all kinds of mystical concepts? An unconscious immanent being is not just a being, it is the Universe itself, and we should lose this ultimate anthropomorphic attitude that children naturally have about things surrounding them. For them, every object has a mind of its own and this phenomenon is related to animism, which defines the attitude of assigning a mind to objects. For our ancestors, the mountain, the river and the thunder were beings. Some have now applied this to the Universe itself, but wouldn't this be going backward? Would it not be best to keep things as simple and rational as possible? The word Universe adequately meets all the criteria for simplicity and is open to research.

But even for the sake of argument, it is inconceivable to imagine an infinite "body" conscious of itself.

Consciousness is a difficult concept to grasp and it implies, at the very least, an awareness of all the parts that constitute the entity. It also implies a kind of information processing of the object of this awareness – in other words, everything that exists. However, infinity excludes such a process. A conscious body must be in a finite body otherwise it has no way of connecting, processing and knowing what would be its own infinite body.

The only way the Universe can be aware of itself is by way of a partial awareness that is manifested through finite bodies, themselves conscious of parts of their environment (inside and outside themselves) but yet found in all parts of the universe. This is precisely what all sentient beings in the Universe are.

We, ourselves, are pieces of the Universe. The Universe is therefore expressing itself through us. Infinity can only become aware of itself through beings who are part of it everywhere and who are conscious of this fact. When we, conscious beings, look at the stars, it is in essence the Universe contemplating itself. We, in fact, are part of the infinite Universe trying to become aware of itself.[24]

---

[24] Well… for some...

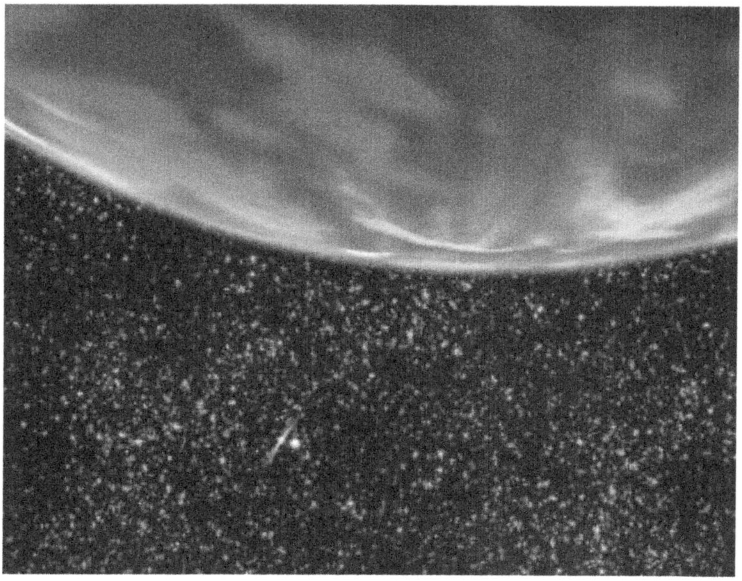

The dark night sky paradox

According to Olbers' paradox (dark night sky paradox), the night sky of an infinite universe should be bright because we should see stars everywhere. In an infinite universe, there are an infinite number of stars, and they cover the whole night sky and would, according to the idea, result in the night being totally bright.

In 1995, the Hubble Space Telescope allowed scientists to observe a very small, completely dark patch of the night sky over the course of 10 days. The resulting images were very humbling; the astronomers could see around 3,000 galaxies that had never been seen before.

As we will show in a moment, this will bring us to believe that the way in which the argument of the dark night sky paradox was constructed was completely wrong. The significant question here is: Why can we not see stars, and even galaxies for that matter, that are really there in the dark patches of the night sky? Whether the Universe is finite or infinite is irrelevant to this question. The answer, of course, is simple. They are too far away from us to be seen with the naked eye. Indeed, the farther away a dot of light is, the smaller it becomes. The number of photons and the density of radiation emitted by a particular star are limited and finite. As the light increases its distance from the target (our eyes), the density of radiation decreases and the photons become sparser according to the inverse square law. There is, therefore, a critical distance where the scarcity of photons is too great and no photons reach our eyes (or a sensor apparatus like a telescope) in a short time. We can perceive a little more light only when a camera is used to record the accumulated photonic hits over a longer period of time (duration of exposure).

If we drew a sphere around Earth, where the radius corresponded to the critical distance beyond which we cannot see the light of a star with our eyes, we would observe something critical: seen from the center, there are not enough stars to fill the inside of the theoretical sphere to entirely whiten the sky. This is why the sky is mostly dark, and why the bright sky paradox is incorrect.

There is, in fact, no paradox. Thanks to the Hubble Telescope experiments, it has been established that we cannot see distant bright objects unless we increase the exposure time of a light (or some other radiations) recording device. This is true whether the Universe is finite or infinite. Therefore this argument cannot be used in a debate about the infinite or finite nature of the Universe.

Zeno's paradoxes

Zeno posited a number of paradoxes with respect to movement that have perhaps not, to date, been satisfactorily resolved. But it is possible to do so using the concept of infinity.

Essentially, all of Zeno's paradoxes are variations of the same principle. He believed everything is one and that there is no plurality. He also believed that motion is nothing but an illusion.

Firstly, we agree that everything is one, but this does not mean that this "one" has to be uniform everywhere. The anti-thesis of absolute emptiness is equivalent to absolute fullness, and since everything is interconnected, infinity is one, but in different densities and movements.

We will use his "dichotomy paradox" to illustrate an approach that does effectively allow us to resolve the paradoxes. The dichotomy paradox attempts to demonstrate that it is impossible to move from point A to point B. Zeno stated that in order to go from point A to point B, you had to walk half the distance, then half of what was left, then half of what was left, etc. But this meant that you would never actually reach point B, because in an endless or infinite loop, you would constantly have to cut the remaining distance by half.

However, this argument only proves infinity. It does not prove that movement is an illusion. What Zeno described was nothing more than a "zoom in" movement, taking us to the infinitely small. This is similar to what happens when you look into a camera and zoom in to enlarge the view or when you use a microscope and zoom in on an object to view its molecular structure. As you can see, this is entirely different from the movement you would normally do when you go from point A to point B. When you walk, you do not zoom into the infinitely small, but simply move at the same level or dimension of infinity. This is why we do move from a point A to B.

Consequently, there is no paradox. Zeno's paradox is itself an illusion that further confuses people into mixing two types of actions: zooming in and simply moving. But once understood, the idea clearly supports the concept of infinity.

III – Life

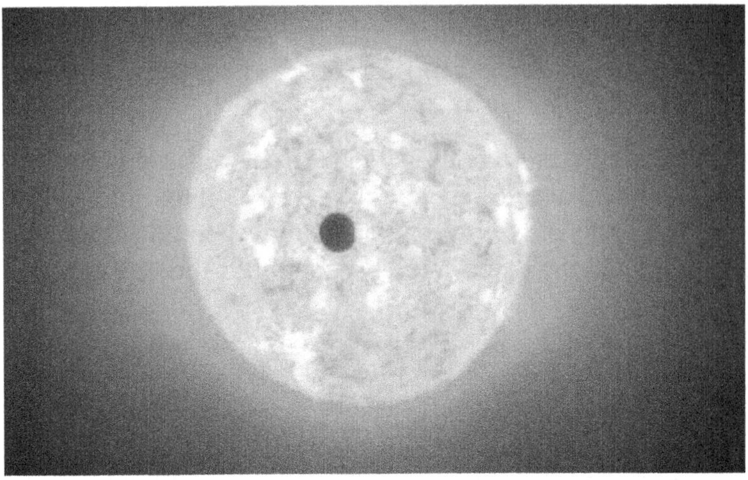

Extraterrestrial life

Most scientists agree that we are most probably not alone in the Universe. Whatever the process by which life originated, it is highly likely that the same course of action took place elsewhere in the Universe because of its infiniteness in space and in time. Even those who believe in the finiteness of the Universe agree on this because this universe is already so vast and "old." With this in mind, it would be difficult to explain how we could be the only occurrence.

Is there a way to estimate how many civilizations there are in an infinite Universe? Logically, we should find an infinite number of civilizations in an infinite universe, but a more relevant question could be: "What is the density of life?" Another one could be: "How many civilizations do we have in our galaxy?" Many people have heard of the "Drake Equation" formulated by Frank Drake in 1960.

However, the problem is that the equation contains seven unknown variables.

$$N = N_* \bullet \int p \bullet n_e \bullet \int l \bullet \int i \bullet \int c \bullet L$$

The formula estimates the number of advanced technical civilizations in our galaxy capable of radio astronomy (N):[25]

N_*, the number of stars in the Milky Way galaxy;

$\int p$, the fraction of stars that have planetary systems;[26]

n_e, the number of planets in a given system that are ecologically suitable for life;

$\int l$, the fraction of otherwise suitable planets on which life actually occurs;

$\int i$, the fraction of inhabited planets on which an intelligent form of life evolves;[27]

$\int c$, the fraction of planets inhabited by intelligent beings on which a communicative technical civilization develops;

---

[25] Loc. cit.: Sagan p. 299.

[26] Thanks to the search for exoplanets, astronomers have found more than 4,000 planets in other solar systems (2019). They have also used the Kepler telescope to find those that have a size comparable to our Earth. Approximately 361 have been confirmed to be in a habitable zone. Exoplanets are now viewed as very common objects in the Universe (http://kepler.nasa.gov/).

[27] This term assumes that Darwinian evolution is the only way life can develop on a planet. This assumption greatly decreases the final estimated number. But Fermi has introduced an idea that increases the odds tremendously.

$\int L$, the fraction of a planetary lifetime during which a technical civilization exists.

This is a guessing game based on assumptions related to a model of the Universe and the origin of life. For one, the formula is based on a finite universe, which makes it obsolete. Indeed, a finite universe implies theories that limit space travel, which in turn reduces the possibility of our planet being visited by extraterrestrials. Because we have established that we live in an infinite universe, this should allow us ultimately to make a better estimate of the average density of life in the Universe.

Given the infinite nature of the Universe, which allows for unlimited scientific progress, it is fascinating to find out where this will lead in terms of the multiplicity of intelligent life forms. Infinite scientific progress would allow space travel at speeds that, theoretically, would increase without any limit. Indeed, as we indicated earlier in Chapter II, speed has no theoretical limit in an infinite universe! The concept of infinity opens up a completely different view of the Universe from the one with which we are familiar (finite universe). We are entering a new paradigm that is much more open than the current one. The idea that we are probably not alone in the Universe makes sense, particularly in an infinite universe (time and space). Not only that, but since the infancy period of a given humanity should be very short compared to the adulthood period, it would be more likely that the extraterrestrials have more chance to be older than us and thus way more advanced than us given that the progression of science is algorithmic or exponential.

The existence of a universe that has the property of being infinite in space and time leads to the following two consequences: (1) as stated, the speed of light is not a limit to speed and velocity exists on an infinite continuum of velocities; and (2) the technological progress will forever continue. Indeed, we have noticed that our technological progress is ongoing and is accelerating.

> Our forebears expected the future to be pretty much like their present, which had been pretty much like their past. Although exponential trends did exist a thousand years ago, they were at that very early stage where an exponential trend is so flat that it looks like no trend at all. So their lack of expectations was largely fulfilled. Today, in accordance with the common wisdom, everyone expects continuous technological progress and the social repercussions that follow. But the future will be far more surprising than most observers realize: Few have truly internalized the

> implications of the fact that the rate of change itself is accelerating.[28]

At some point in the future the acceleration of progress will change drastically and will be incredibly steep. Science philosophers call it the *singularity*.[29] In an infinite universe, the trend of progress does not and should not have any limit since there is an infinite range of particles with their unique vibrations and an infinite pattern of interaction between these particles, and this is for the humanities to gradually and forever discover.

These two points lead to an infinite continuum of speeds in terms of space travels. The infinite universe opens up a completely different perspective relative to our capacity to reach the stars with our future space ships. Our galaxy becomes within our reach, contrary to the current view where this would seem to be impossible given the estimate that our galaxy is around 100,000 light years in diameter. Since our progress will be infinite and constantly accelerating, we can forecast that our civilization will eventually reach a point where it will be quite easy to one day travel through it relatively quickly and reach other solar systems. This, in turn, will enable us to create and propagate life on other planets much faster. If we simply take a look back at our own history and our track record of colonizing continents, we can conclude that it is part of our nature to propagate and "spread our seed." It seems to be inherently a part of our biological makeup and represents the most likely scenario as our scientists already have plans

---

[28] Kurzweil, Ray, *The Law of Accelerating Return*, published on KurzweilAI.net, 2001.

[29] Kurzweil, Ray, *The Singularity is Near*, Penguin Group, 2005.

to establish life on Mars. We are just in the beginning stages of space exploration.

> We have a moral obligation to plan for the propagation of life, and even the transfer of human life to other solar systems… Securing that future for life can give our human existence a cosmic purpose.[30]

When our scientists master the ability to create life artificially, they will also be able to create life forms with the unique ability to adapt to the environment of planets completely different from our own. There is a high likelihood that this mastery will occur and would probably be the most practical method of propagating life on other planets. This also will allow for a more flexible and efficient process of propagation and will mean that the speed with which life proliferates will greatly benefit from knowledge of genetic engineering.

> Consider just some of the progress associated with the J. Craig Venter Institute. In 2003 its researchers created a synthetic version of the bacteriophage phiX174. In 2007 they successfully transformed one species of bacteria to another by genome transplantation. Most recently, they developed methods for the complete synthetic assembly of the genome of the bacterium *Mycoplasma genitallium*. The techniques now developed make the feat of sequencing the human genome in 2001 seem prehistoric. Not only have the cost and speed of sequencing evolved faster than those of computer chips, but the ability to use both chemistry

---

[30] Interview with Michael Mautner on PhysOrg.com about an article that will be published in the *Journal of Cosmology* (March 2010), www.physorg.com/news184915200.htm

> and biology to synthesize new complex organisms has undergone a revolution in the past five years.[31]

The Craig Venter Institute reported in May 2010 on the creation of the first bacteria cell controlled by a chemically synthesized genome.[32] This made big news in popular media and questioned the origin of life in a powerful way.

> Atheists of the Dawkins type will take it as practical proof that there is no need to hypothesize God at all: we can make life without any miracles, and there's no need to imagine a creator.[33]

The term "creator" in this excerpt refers to a supernatural god, but the ability to synthesize a functional DNA strand also makes us creators, albeit scientific ones.

When we start propagating life in the Universe, we will also make progress with respect to the process of creating life. After the scientific creation of a few artificial life forms on other planets, the process could be automated. Overall, this process would take the shape of an exponential curve, similar to the propagation of life on our

---

[31] Krauss, Lawrence M., *Life from a Test Tube? The Real Promise of Synthetic Biology*, from the February 2010 Scientific American Magazine: www.scientificamerican.com/article.cfm?id=the-real-promise-of-synthetic-biology

[32] Venter, Craig et al., *Creation of a Bacterial cell Controlled by a Chemically Synthesized Genome*, Science DOI: 10.1126/Science.1190719, 20 May 2010 (http://www.sciencemag.org/cgi/content/abstract/science.1190719).

[33] Browns, Andrew, Has Venter Made us God?, www.guardian.co.uk, 20 May 2010, http://www.guardian.co.uk/commentisfree/andrewbrown/2010/may/20/craig-venter-life-god

own planet. This suggests that out of one civilization, we would obtain at least two civilizations in a few tens of thousands of years (something between 20,000 and 30,000 years).[34] We could then expect a further expansion in the order of four in a subsequently shorter period, then eight, then 16, then 32, etc. Assuming the length of time for each step in the process of propagation is the same (which is a conservative notion since we should expect automation to speed up the process), in roughly one million years that process would have created enough civilizations to populate an entire galaxy the size of our own by simply placing one civilization within each solar system! In about two million years, the "known" universe would be full of life. This means that every solar system in every galaxy of the "known" universe would contain a civilization. This is how powerful this process is. It should be noted that the automation process would balance out the loss of civilizations that would destroy themselves through nuclear wars were they not able to control their own violence and aggression. But also, the planets of self-destroyed civilizations would be recycled to put life back on them.

All of these created forms of life would have originated from the deliberate and scientific creation of their creators and none of them would have originated from "natural" chemical and biological evolutionary processes. In fact, in such a context, these "natural" evolutionary processes

---

[34] Using our own planet's estimated population growth, we can infer, using an inferential statistical model, that it would take between 20,000 and 25,000 years to obtain a civilization of seven billion people on a new planet, starting with only seven couples. Therefore, since a planet's population correlates with the level of scientific progress, one can predict the other.

would not have much chance of occurring on planets. Indeed, the intelligent propagating process would cancel it out by putting life relatively very quickly on available planets. An intelligent propagation process would take one million years to populate our own Milky Way entirely while it is believed that it would take 3.5 billion years for life to evolve (chemical and biological natural type of evolution) on just one planet and not necessarily in our own galaxy.

All of this suggests that if we travel into space and meet extraterrestrials, the chances of them having originated from some sort of seeding or a complete scientific creation are statistically highly likely, and would be of the order of 99.9999999999999999999999999...%.[35]

Most importantly, and as a boomerang effect, this prediction also suggests that our own civilization's origin could be part of an infinite continuum of creation of life from planets to planets, i.e. part of an extraterrestrial civilization's life propagation project. In other words, the conclusion of our prediction contains the explanation of our own origins.

---

[35] If one compares the number of humanities created scientifically with the one that would be generated from random chemical and biological evolutionary processes, as proposed by Neo-Darwinism, over a 3.5 billion year period, the ratio of the two numbers would give an idea of the probability scale on which the former is most likely to occur compared to the latter. The probability of finding another civilization originating from a scientific creation process would be in the order of 99.99... percent with tens of thousands of 9s after the decimal. Would this not be equivalent to a certainty?

In brief, if we can predict that we will create life on other planets, it suggests we may already be part of an ongoing life propagation process. Furthermore, unless we are the first civilization to appear in this part of the Universe (that is, if we believe that life forms can originate from purely random chemical and biological processes, AND that we are the first one to appear that way in the universe), it is extremely likely (more a certainty) that our origin is linked to a propagating process that involves a more advanced civilization, our parents from space. Since our civilization has not been advanced enough in the past to understand this basic principle, scientists have developed other theories to explain our origins, theories that are actually part of our socially accepted paradigms.

But if our origin is linked to a type of organized and directed "panspermia" by beings from elsewhere, then "where are they?" This famous question was asked by Italian physicist and Nobel Laureate, Enrico Fermi.[36] He held the same reasoning, which was brilliant considering that he raised this very question at the beginning of the 1940s. At that time, the UFO phenomenon was relatively unknown. However, a few years later, in 1948, this phenomenon had become common knowledge, especially when the first modern UFO sightings were publicized and provided a glimpse into the answer to Fermi's famous question.

The disinformation promoted by the various governmental agencies influences scientists even today into rejecting the reality of this phenomenon. It is an easy task, though, as the

---

[36] Crick, Francis, *Life Itself – Its Origin and Nature*, Simon & Schuster, New York, p. 13.

accepted paradigm in physics considers extraterrestrial visitations highly improbable. This paradigm is so powerful and so deeply entrenched that most scientists would not even consider investigating the UFO phenomenon.[37] This also explains why Fermi's proposition is commonly known as "Fermi's paradox." Based on the math of life propagation, scientists understand that we should most likely have been visited by extraterrestrials (actually life on Earth would have been seeded by extraterrestrials – the terraforming concept), but the accepted paradigm forces them to think otherwise. Therein resides the paradox. Instead, the scientific community awaits extraordinary evidence because the paradigm they believe in requires it. But all the evidence that exists is not really looked at. It is inconvenient.

This partly explains that while we, our societies, should expect UFO visitations according to Fermi's proposition, we are very quick to dismiss all UFO reports and, at best, when they indicate a more-advanced-than-us visitation, classify them as "unexplained phenomena." The fact that these reports are unexplained serves the purpose of avoiding concluding in a positive light all matters related to the UFO phenomenon. It serves to show that no evidence that can potentially justify a change of paradigm will be accepted. Another way that has been used to stop the existing investigations about UFOs is to conclude that they represent no threat to the public. Indeed, "no threat" does

[37] In fact, the scientific community, as a whole, has become paralyzed at a stage of complete denial just as the paradigm with regard to extraterrestrial visitations has begun to change in the mind of the population because of the increase in strong UFO-related testimonies and the associated evidence.

not mean interplanetary UFOs do not exist... but it does answer the question that started the investigations (are UFOs a threat to national security?). Partial answers work wonderfully in politics. It really looks like there is a cover-up done by the military that is very convenient to the scientific community. The best scientists are probably in the cover-up and help the cover-up. Disinformation is a tactic used by the military to muddy the water about the UFO phenomenon. And the sceptics of all fields of science are having the fun of their live. But this type of skepticism is counterproductive and part of this effort is intended by the cover-up. The other sceptics are just following the wave.

People usually believe what they want to believe no matter what takes place around them[38] and are blind to anything that threatens or goes against their belief system, even if it occurs in plain view. In addition, a strong factor in the lack of reactions from the scientists is the fear of losing their reputation, being excluded or ostracized by their peers, which often results in job loss or absence of funding.

We are always amazed by people who say there is no proof of UFOs. It just shows that they just believe the propaganda on this topic. Valid and very credible information is out there and it is one's duty to find it. We are not going to investigate the UFO phenomenon for the readers. If one opens one's mind to it and takes the time to research the subject of UFOs, one will undoubtedly find that some very

---

[38] This book will undoubtedly be largely dismissed as it represents a complete paradigm shift when compared to the current paradigm.

strong evidence exists.[39] But there is a lot of effort out there to discredit the information and it is up to each of us to discriminate between what is true and what is false.

There also seems to be traces left here and there in our distant past such as ancient descriptions of beings descending from the sky telling us that they created men "in their image." It looks as if we invented the gods when we saw *them* coming from the sky as described in ancient writings.

The natural tendency of primitive humans is to be anthropomorphic. This produces animism where the psychological reflex leads us to believe that objects, whether plants, rivers, thunder or volcanoes, have a human mind. Children act in exactly the same way, thereby demonstrating how natural this human tendency is. The point here is that this is quite different from deistic religions.

Believing in beings that came from the sky and created all life on Earth is so unnatural[40] that it suggests an entirely

---

[39] Just one example: In 1990, General De Brouwer of the Belgian Air Force held a press conference to show the efforts of the military to intercept UFOs. Radar recordings were shown. For De Brouwer's official statement, google "Wilfried de Brouwer": This not only shows that the evidence exists but also indicates one of the most common reactions of a society facing the perception of the "impossible," i.e. nothing is done. This episode is a rare exception, and a cover-up is usually what occurs, especially in the USA where only a small group of "powerful" people wants to have total control over this situation. More recently, there was the Nimitz UFO incident where pilots encountered UFOs. This was confirmed reluctantly by the CIA.

different process. By this we mean that it appears as though deistic religions may have originated from contacts that our ancestors had with extraterrestrials and they were so impressed that they recorded these otherwise incredible events by way of their own related rituals.

Historical records provide evidence of this sort of phenomenon. When Americans landed on the island of Tanna during World War II and met the Tannese people, this population had had, up until that point, no real contact with a technological civilization. The Tannese were so impressed by the Americans' technology (jeeps, helicopters, guns, machines, etc.) that they perceived them as "divine," adhering to the notable third law of Sir Arthur C. Clarke:

> Any sufficiently advanced technology is indistinguishable from magic.[41]

From being animistic, they became deistic, as they perceived one of the Americans to be a god, and thus began the John Frum Cargo Cult. What happened on Tanna Island could have happened thousands of years ago when our ancestors were contacted by those who propagated life on Earth and purposely left messages about their creation (ancient texts and scriptures) for us to understand when our minds had acquired enough knowledge to do so.

---

[40] Sociologists were forced to assume that the origination of deistic religions could have come only from the imagination of humans as if there were no other possible scientific explanation.

[41] Clarke, A.C., "Hazards of Prophecy: The Failure of Imagination" in the collection Profiles of the Future: An Enquiry into the Limits of the Possible (1962, rev. 1973), p. 36.

It is fascinating that the process of propagation of intelligent life that is suggested here is actually supported by deistic religions, which describe messages from beings who came from the sky through descriptions that precisely fit this proposition. The process of propagation of intelligent life suggests that we should expect deistic religions to develop as our ancestors encountered their creators and receive their message, and this is exactly what we find.

Another Nobel Prize winner, Francis Crick, used a comparable life propagation argument and inferred that primitive forms of life were seeded on Earth by automatic spaceships of extraterrestrials.[42] This process of propagation is so powerful compared to the current theoretical process of the genesis of life[43] that it becomes the most probable. We should expect extraterrestrial visitations to occur not only in the present but also to have occurred since the beginning of our civilization and even earlier. We mean by this that intelligent life forms would explore every potentially suitable planet on which to propagate life and this process of propagation would, therefore, impose itself as the major, if not the only, process of propagation of life because of its enormous relative speed.

The immense power of spreading life throughout the universe by way of a scientific creation of life suggests that we most likely originated through the intervention of extraterrestrials. This assertion is supported by the UFO

---

[42] Loc. cit.

[43] Theories of chemical evolution.

phenomenon and would translate to our creators (or extraterrestrial neighbors) being able to observe our progress, as well as religions describing encounters with beings who came from the sky with a message about life's creation.

Intelligent Design

I think we should open our minds to the idea of creation, not from a religious point of view, but from a scientific perspective. We have shown that the existence of a God creator of the universe would lead to a paradox. Evidence of a design for life would support more rationally an extraterrestrial intervention on Earth in the distant past. The existence of deistic religions can be perceived as evidence of a controlled panspermia, especially if the message from the alleged *divine* beings were related to the creation of life on Earth.

The current scientific paradigm holds that life appeared through chemical evolution from inorganic matter. There is, however, a problem with this idea and it motivated Francis Crick in his book *Life Itself*, in which he supported the idea of an intervention by extraterrestrials rather than a chemical evolution as the origins of life. This is perfectly in line with what we are proposing here: a directed panspermia process, which is more likely to have occurred. This deserves serious consideration. Not only is controlled propagation of life highly probable, but also its alternative theoretical competitor, chemical evolution, is highly improbable.

The question we should ask about the chemical evolution hypothesis regarding the origins of life on Earth is how random events could have created the information needed to sustain a living organism and reproduction system – information that is then interpreted by a would-be primitive cell to function in all its aspects. To claim that chemical natural selection played a role is completely irrelevant because of the coding. DNA or any other form of molecular coding has to appear before any mechanism of natural selection can play a role. Furthermore, such an entire system has to be randomly assembled in an instant to stand a chance of continued existence, otherwise an incomplete system will not survive. This scenario is so highly improbable that the belief in random assembly becomes a kind of myth.

The old Miller-Urey experiment and others that followed are trying to provide evidence that life can emerge from inorganic matter, a sort of chemical evolution on the young Earth. However, the important point here is that the amino acid molecules that were randomly created in the experiments do not represent a system that can carry information. We find all kinds of molecules in living organisms but not all of them carry information and this represents the core of the debate. Without carrying information, the first assembly of "life molecules" is utterly useless and cannot lead to life. Life forms are nano-machines that are self-replicating, self-repairing, etc. These processes need information systems that are independent of their support while remaining in connection with the reality of the life form that the information system encodes as well as its environment. This is the essential problem for believers in random-accident miracles. Evolutionists still believe in the possibility of random accidents creating life in a gradual way but scientific experiments have failed to support this view.[44]

Mathematicians looked into the problem many years ago[45] and to this day have not come up with any strong arguments showing that random[46] processes can create information. The myth of evolutionists is that with an infinite number of trials and errors, you find a "solution" to any complexity.

The Universe has laws and all the parts and constituents have to obey them. These laws explain the natural influences upon atoms and molecules and they do not have a choice but to be assembled according to specific laws of chemistry and physics. A more simplistic way to look at this is to say that atoms and molecules follow the easiest path directed by the laws of physics and chemistry. This means that the laws of the universe limit the space of all possible events and it seems that random simulations have not accounted for this limitation.

---

[44] An example: Vera Vasasa, Eörs Szathmáry and Mauro Santosa. Lack of evolvability in self-sustaining autocatalytic networks: A constraint on the metabolism-first path to the origin of life. *PNAS*, 4 January 2010, DOI: 10.1073/pnas.0912628107, cited in Denton, Michael, Evolution: A theory in Crisis.

[45] Moorhead, P.S. and Kaplan, M.M., eds (1967), *Mathematical Challenges to the Darwinian Interpretation of Evolution*, Winstar Institute Symposium Monograph.

[46] In reality, we refer to pseudo randomness. Pure randomness makes no sense whatsoever because there are always reactions from preceding reactions extending in an infinite time continuum. The appearance of randomness is due to the ignorance of the causes. But what statisticians mean by randomness is a lack of correlation between events, which, in turn, implies uncorrelated causes.

Why should we believe that "the easiest path" of the chemical laws correlates with the creation of information? A correlation needs an explanation else it hides an unknown principle. It is not science anymore; it is magic. But even without taking this factor into account, realistic computer generated simulations[47] using randomness as an algorithm have failed to demonstrate any creation of information. Let us repeat that gradual chemical evolution would not work until a completely well-functioning system was achieved, otherwise the organism itself could not survive. If gradual chemical evolution was possible, it would need to have everything built in to survive and reproduce itself, otherwise it is totally futile. And this would not be a gradual process. This further exceedingly decreases the probability of a chemical evolution.

Simply put, poems need poets while a random distribution of letters on a piece of paper would be merely a collection of symbols without any information and meaning. There is no evidence whatsoever that random events can create a coding language on the level that is needed to create life.[48]

---

[47] The minute quantity of information that a given simulation could generate is always guided by an intelligent algorithm (when trying to simulate changes and hoping to find information in the result). This contradicts the primary engine of any evolution theory, which is randomness (in fact, it should be guided by the laws of the least effort as some movements are easier than other). Secondly, a complete information system is needed to keep track of the changes (when trying to simulate natural selection), which is not possible if the approach is to simulate the genesis of an information system. It is a catch-22, the very paradox of evolutionists.

[48] Dean H. Kenyon co-wrote *Biochemical Predestination* (1969), a book that was a reference on how life could have originated from natural chemical evolution. He discovered in the years that followed the

To believe otherwise is faith, i.e. it is not supported by evidence. The evidence shows the opposite. It looks like the god of the evolutionists is "pure chance." It is a type of mysticism that the evolutionists themselves are not even aware of.

On the other hand, the alternative theory for the origins of life should be scientific. It should be part of what exists under the known laws of the Universe. If life, because of its informational structure, cannot originate from inorganic matter ruled by the laws of physics and chemistry then it must originate from the artificial domain. Life must be artificial by the very nature of its structure, which is not due to natural causes! Life is essentially based on information carried by DNA coding and this cannot have self-generated. Because we know for certain that only a higher intelligence can create information, it follows that the information contained in life was also created by intelligent beings. If it was not us, it was *the others.*[49]

The problem is not that the basis of the deistic argument rests on the body of evidence that relates to the theory of Intelligent Design (ID) or creationism as their favorite weapon to bring the ID discussion into schools; rather, it is that they hope the concept of God will be revitalized. They fail to realize that the only source of design admissible in

---

publication of his book that his own theory was flawed and based on wrong assumptions. As new facts were discovered, he realized that life could not have originated from natural chemical evolution. He has since then joined the Intelligent Design movement.

[49] *The others:* Curiously (or not), it is also one of the translations of the Hebrew word *Elohim* found in the Bible and identifying the *beings who have come from the sky*, yet an alternative translation.

science must rest on a rational cause, not a vague and concept such as God. Hence, the only option left is for this design to have come from intelligent biological entities, an extraterrestrial civilization.

What a turnabout for religious activists! They do not realize that they are bringing about the beginning of the end of their movement. Nevertheless, they serve a purpose for the moment as they have uncovered scientific and rational evidence of a design in living things. This can be used to support the idea of an extraterrestrial intervention in the creation of life, thereby reinforcing the idea that the supernatural does not exist and removing the only scientific support that deists thought they had. This is why the Theory of Intelligent Design (ID) is the most powerful weapon that atheists have in the fight for the removal of mysticism in the search for the genesis of life.

The renown atheist and evolutionist Richard Dawkins was interviewed by Ben Stein for his movie *Expelled: No Intelligence Allowed*, and was asked a question about the possible eventuality that the Intelligent Design theory rather than evolution may turn out to answer some of the issues in genetics. The following excerpt contains his answer verbatim.

> …it could be that at some earlier time, somewhere in the universe, a civilization evolved by some kind of Darwinian means to a very high level of technology and designed a form a life that they seeded onto perhaps this planet. That is a possibility, an intriguing possibility, and I suppose that you might find evidence for that if you look at the [cell?], biochemistry, microbiology, you might find a signature of some sort of designer… and that designer could well be a higher intelligence from elsewhere in the universe. That higher intelligence would itself have had to have come

> about by some explicable… or ultimately explicable process. It couldn't have just jumped into existence spontaneously, that's the point.

What Dawkins wanted to show is that even if the facts were to show that life was intelligently designed, which is contrary to his position, the source of this design would certainly not be a God. As Dawkins did, scientists in all fields should come to the realization that the Intelligent Design theory or other arguments proposed by creationists are not real threats to science. Intelligent Design theory will never impose the idea of a God. It will rather suggest the idea of an extraterrestrial intervention by biological entities, an idea that is void of mysticism. This argument should be used widely in order to show the irrationality of the equation Intelligent Design = God! Instead this equation should be Intelligent Design = extraterrestrial intelligent biological entities.

On the other hand, scientists have ridiculed all ideas of creation because they believed that this perspective carried the concept of God. In doing so, they have "thrown the baby out with the bath water." Scientists have been biased in selecting only the interpretations that did not suggest a design. By rejecting the idea of a creation altogether, they also developed a blind spot regarding the idea of a creation by extraterrestrials. But it just happens that it is the very idea that can really help us understand our origins and destroy the link that deists make between the scientific argument of a design and the concept of God creator of life.

Not all the arguments of creationism and Intelligent Design theory are wrong. Some are quite interesting and could be used to support an alternative scientific theory involving a design carried out by extraterrestrials. Simply considering the possibility that scientists from another planet may have come to Earth to perform scientific experiments using genetic engineering will help eliminate the idea of an Almighty God in a more powerful way.

Interestingly, people generally say that there is no proof that extraterrestrials have left traces on Earth when there is plenty of visible evidence.[50] Beside the evidence of the existence of religion testifying that beings came from the sky and created life on earth, another type of evidence stares us in the face and that is life itself. Life is actually an artifact left by extraterrestrials but people cannot see it. We labeled life as "natural" but it bears the mark of artificiality. Life is a technology! It is a nanotechnology that surpasses our own technology, but we are so indoctrinated to view life as natural that its artificial character goes right over our heads, especially the heads of evolutionists.

My revolution in researching the origin of life

I remember when I first learned about the idea of extraterrestrials creating life on Earth. I was rather surprised that I had not thought about it as a possibility sooner. I was an evolutionist at the time and I had already investigated the UFO phenomenon, resulting in a positive conclusion. I embraced the theory of the ancient astronauts who landed on Earth a long time ago and changed or intervened in the process of our natural biological evolution. However, this new idea of a complete creation

[50] The evidence is automatically dismissed or ignored by the scientific community because it is not part of the current paradigm. They always find a natural way to explain it even if it is very unlikely or they simply ignore it altogether.

(without evolution) needed to be investigated. I needed to see if it held water. The psychological trick I used was to play devil's advocate against my own view as if I were playing chess with myself and changing sides after each move. Being an evolutionist, I gave myself a challenge: to find evidence of a scientific creation that I would not be able to explain using the evolution paradigm, which was my belief. I forced myself to think as an atheist creationist to help myself in finding clues against my own belief. Doing this changed something in me. I became more objective and I perceived facts and arguments that I had not paid attention to before.

I started to pay more attention to shapes and colors, noticing, for instance, the distribution of colors on the scales of tropical fish and on the feathers of tropical birds. The colors did not appear as though they were randomly distributed at all. They looked rather like a work of art.

A line of color that crosses the many feathers of a bird needs a very precise and calculated DNA programming. Each part of the line is drawn at a different location on each feather! The understanding of what it takes in the DNA to accomplish this excludes any interpretation leading to randomness. I could not explain this phenomenon through random mutations and natural selection, which are the twin engines of evolution. I do not think they can. Without providing any solid explanations, they simply believe that the evolutionary process did it somehow because evolution is essentially a belief! Sometimes, they give a pseudo-explanation to get it out of the way. For instance, they would say a male with more precise colored lines would be more likely to be selected by a female and reproduce better, while all other males would not be selected at all. This is an illusion. It does not explain how the line was originally assembled. Besides, a creator could program the DNA to obtain this natural selection mechanism, encouraging his created species to minimize the deterioration of the DNA code through accidental mutations. The main problem with the evolutionistic rationale for all the changes that would have occurred is that it does not explain how the DNA was programmed in the first place. Natural selection can only work on the DNA that is already there. Instead, natural selection could simply be seen as a mechanism of protection of the DNA code.

One day I found a dead monarch butterfly, intact, by the side of the road. When I looked at the fine black lines drawn on the wings to highlight the precision in the arrangement of the small scales that provides the colors of the wings, I was astonished to see that the black lines were mostly distributed on top of the fine rigid structure of the wings.[51] Structurally, this frame consists of tiny tubes called "veins" to which the tissue of the wings is attached so that the wings remain straight when flapping. This is a required condition for flying. The presence of these black lines surprised me because they showed an even more improbable event than the obvious designed of the distribution of colors found on butterfly wings. I thought of the complexity of the DNA programming it must have required distributing these tiny black scales in such a fashion as if to precisely hide the veins of the wings. Also, it occurred to me that this is what a designer artist would do. I could not imagine or find in the literature any other cause except a designer's conception to explain such a highly improbable correlation. I calculated that the probability of such an event occurring would be in the range of one chance in a number containing more or less one million zeros ($1/10^{1,000,000}$).[52] That simply puts a number to what we can already intuitively feel. This number is so big that no normal book can contain it (nearly 500,000 pages of a book this size). It means that such a correlation cannot have randomly happened. Even to code a simple colored line, no matter where it is on the wings, is an improbable event because a line is a correlation between the colored scales and their location and it needs very specific and calculated DNA programming. Not only can we see lines on the wings of the butterfly but many are on top of the veins and these should be two completely

---

[51] I did not notice this before because I was wearing the glasses of an evolutionist. I was therefore biased.

[52] I invite the statisticians to do the same, and no matter what method you use, the difference in results will not change the conclusion.

unrelated structures (veins and scales)! I was stunned! Life is showing us a design, but because we have been educated to think as evolutionists, we ignore the depth of the role that this design means in living entities. This evolutionary paradigm also prevents us from realizing the high degree of improbability that the information contained in living entities could have been randomly generated.

I realized at that point that the idea of evolution in which I had believed concealed a central mysticism: It hid the belief in the magical powers of randomness. Little by little, I realized that the genesis of life was better explained by a scientific creation, whereby the designers would be as human as we are. This explanation also satisfied my view, which needs no mysticism whatsoever.

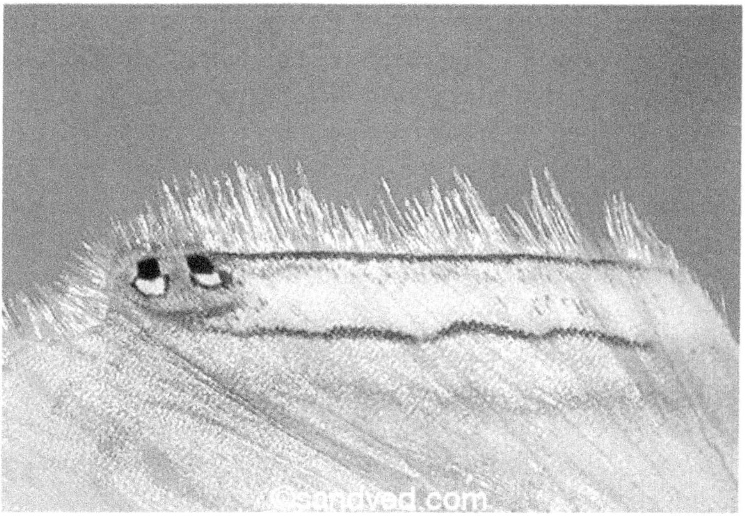

Images from Sandved.[53]

---

[53] For some reason, the particular butterfly picture of the human face worm drawn on a butterfly wing found above was removed from the website www.sandved.com after Sandved accepted that I use it (Sandved adheres to the evolution theory). But this is a very cute artifact if it really exists. If so, it is obviously a veiled message from those who designed it… Many other photos of designs found on butterfly wings can be found on the website http://www.butterflyalphabet.com, which leads to the same conclusion. I have placed the interesting ones here for convenience:

https://docs.google.com/document/d/1ivUUHLjGcXShrqPiL_c5pXQT9fkSH5whRFrI3FH4h8U/edit?usp=sharing (or google search for #sandvedbutterfliesselection. Also, we find very cute designs in the orchid family. Here is an image summary of interesting orchids:

https://drive.google.com/file/d/1h8_xK7vQhqNcqT-WLpt-VZoBhE6WU9Cd/view?usp=sharing (or google search

#orchidsbydesign)

Sandved is a photographer who took pictures of designs on butterfly wings. Like his fellow evolutionists, he advanced the argument that everything we see on butterfly wings comes from mutations and natural selection. But a designer would also design what we see on butterfly wings and, for instance, design the protecting mechanisms for the butterflies they engineer, such as an image of a caterpillar drawn on the wingtips, as illustrated above. A bird will try to get the fake caterpillar and miss the body of the butterfly, which will then fly away. A designer can plan for that. But none of the evolutionary theories explain how the design was put there in the first place. Average evolutionists believe in the illusion that natural selection is somehow acting on the genetic mutations, which cannot be.

Mutations have to occur before they are selected. Other designs on butterflies' wings would also represent images that the artists would have in mind from their own world or the new world created on Earth. When this design illustrates a specific recognizable object or animal, this indicates the artistically artificial nature of the design.

Random mutations have decidedly miraculous powers in the minds of evolutionists because the designs we find in butterfly wings are highly improbable and practically impossible randomly. Furthermore, you will notice in the above picture that the caterpillar design on the butterfly wing has a friendly human face: eyes, eyelashes and a large smile! This is precisely what we should expect from a designer (even a human designer), while it is a highly improbable event from the evolution paradigm's point of view. Other designs found in butterflies are similar. The level of improbability of finding them is so high[54] that considering that it could have happened through a random process defies logic.

For a long time, many facts have been swept under the rug and ignored while others that fit more with the creationist paradigm have been hijacked by evolutionists. It is time to open the door to the idea of life having its origin from being part of a chain of creations by intelligent biological entities. We will, eventually, discover the extent to which we have been biased toward some evolutionist interpretations and how mainstream scientists have handled the facts and scientific publications on this fascinating subject. For instance, it is highly probable that evolutionists have been somewhat biased in selecting only facts, experiments and publications that support the evolutionary theory.[55] It is not a scientific endeavor, but an ideological one.

---

[54] The probability is in the order of one chance in a number that has at least thousands of zeros and this is a very conservative number. What that means is that it is practically impossible.

[55] The wisest of scientists are realizing that the truth is bigger than them; hence integrity with the well-being of humanity should be more important than reputation or research funding. This is a condition for a paradigm shift: A minority needs to make the heroic gesture of saying what they really think so that the paradigm shift occurs. It will occur nonetheless, but who will be the heroes who will participate in this scientific revolution? It is time to choose what role one wants to play in this important turning point in the earth's history…

This virulent fight between creationists and evolutionists, however, is no longer necessary. If there was a creation, it would be a scientific one. Regardless of whether it is creation or evolution, there is no mystical God! The only intelligence in the Universe has to be intelligent biological life forms. If some people say that God is an intelligent life form, then it is not a God.

The origin of life in the universe

There is still a popular question that lingers and remains unanswered: "Was there a first life form that appears in the infinite universe?" Firstly, the life propagation process is not forced to have a beginning because, as we have shown, the universe itself has no beginning. Secondarily, if life cannot originate from natural processes, then we are compelled to conclude that life itself has no beginning in an absolute sense. Life is part of infinite chains of scientific creations, or controlled panspermia. This fits very well with an infinite universe that has no beginning and no end. Theories of the universe that include a beginning have to include a beginning to life, and doing so shows how theories in physics can influence the choice of theories in the genesis of life. If one chooses to agree with the Big Bang theory, one is forced to agree that a chemical evolution must have occurred sometime in the past and then a controlled panspermia process would take over. But if there was no beginning to the universe, life could very well come from an eternal process, especially if the evidence shows that life can only come from an intelligent process. If life is artificial, then it comes from an eternal process. All we have shown here point to that conclusion.

This is a crucial point. The body of evidence from computer-generated simulations and theoretical statistics has shown that information cannot be generated from random events. It comes from artificial processes. That we know. We also know that life uses DNA and this is essentially an information-based system. It is actually the most efficient information-based system known to man and the only one that does not come from the earthling's mind.

Life is a system, however, that we would ultimately have created in the future as a more technologically advanced society.[56] Therefore, life has to belong to the artificial realm, which, in turn, has to come from an endless or infinite process of creations. Why? Because life cannot come from nothing or from a natural process and will always need intelligent life in order to be created and propagated throughout the universe. Because there is no beginning to an artificial cyclic process, this ultimately means that the universe had to be without a beginning. Such a fascinating conclusion is shared independently by the demonstration provided in Chapter II regarding the infinity of the universe. It is an interesting cross-validation point. Indeed, the artificiality of life imposes that the Universe cannot have a beginning while, independently of that, we have shown that the universe can only be infinite.

Life and immateriality

There is something in life that can be seen as immaterial. The essence of life is not the atoms and molecules that compose it. Rather, it is the information contained in the

---

[56] This implies that this DNA technology comes from a more advanced civilization than our own.

DNA and its activation. But information is kind of immaterial. Information is not the molecules that code it. The same information could be supported by many other types of material. Indeed, you can write the same poetry on a piece of paper, digitally in a computer file or carve it on a wooden bench. While information absolutely needs matter to exist, it transcends it.

One day, when we have created human beings on other planets, we will try to explain to them that they have been made from dust (since they would not know about atoms and molecules). When they ask why they do not look like dust, we will tell them that we put our thoughts, our mind or our "spirit" (the knowledge of DNA and its coding) into the dust to make it alive and make them look like they do. That is why they will keep this image in their mind that life contains a *soul* that still exists after death.[57] This is exactly what our own religions have retained from the teaching of the gods who came from the sky. It is all making sense.

---

[57] Indeed, DNA still exists after death in bones for a long while and this information could be used (coupled with another information system that would keep the information about the personality of an individual, the connectome) to recreate the individual that lived through an advanced technique of cloning.

Image of the NASA Spitzer Space Telescope of the Double Helix Nebula.

A message from ETs?

Fermi's basic principle of life propagation implies that the universe is full of life at different stages of civilization since progress in science and technology should be endless (no limits to what intelligent life can do; no limit to speed in space travel). This means that some of the civilizations out there could be advanced enough to manipulate stars and utilize them in a way that is helpful for some advanced purposes…

We need to view space in a different light because, given this new vision, there may certainly be many celestial bodies out there that have been modified by intelligent life forms and therefore could be considered artificial planetary bodies.

Let us imagine intelligent beings that were millions of years more advanced than us. What could they possibly not be able to achieve? How many different layers in the infinitely small and the infinitely large could they control? Would they be able to physically move stars? How much faster than the speed of light would they be able to travel – thousands of times faster?

In our galaxy, these beings would be able to communicate among each other and even meet, based on the idea that space travel would not be limited to speed as we have shown in a precedent chapter. This would require a kind of common astronomical "clock" in order to be able to coordinate any type of planned communication. If we were to find such clocks, they would provide additional support for the abundance of life in the universe and its ability to communicate effectively.

In 1967, a type of rotating star that emits a beam of radiation was observed for the first time. For an observer who happens to be on the beam's path, a pulsation is registered due to the rotation of the star. This is called the "lighthouse effect." These objects are called "pulsars" and their pulses are in the range of milliseconds. The regularity of the pulsations can be compared to those of atomic clocks (even more precise) and suggests that they could be beacons and clocks from extraterrestrials. This is why the first pulsar was nicknamed "LGM" (Little Green Man). Naturally, the "artificiality" idea on this label for the pulsar was tongue-in-cheek given the mainstream paradigm. But in light of the argument that we have presented that supports the infinite nature of the universe, this is exactly what we should expect from a galaxy populated by intelligent entities. It also supports Fermi's idea. Pulsars should be part of the list of artificial celestial objects that exist in the universe. We, ourselves, could use a pulsar to measure "time." When the time comes for us to meet our extraterrestrial "caretakers," they will introduce us to the "galactic clock-pulsars" so we can have scheduled communications and meetings with them.

On 16 March 2007, another discovery was made by US researchers while observing space through NASA's Spitzer Space Telescope. They spotted a nebula shaped like a strand of DNA in our own galaxy[58] some 80 light years long (see above picture). Maybe this nebula has been formed naturally, as anyone might assume, but perhaps it is artificial. Maybe it is a "neon sign" from intelligent beings millions of years more advanced than we are as they are playing with stars and moving them around in an artistic manner. Perhaps they are sending a message to all those who can observe it: "Life is out there!" But maybe it is too much for our brains to accept the reality of such concepts and scientific advances. After all, we are still very primitive and our humanity, which is like a living entity, is not unified yet. It is still a "fetus" period and has not been born yet into a more mature civilization.

---

58 http://www.nasa.gov/vision/universe/starsgalaxies/milkyway_twist.html

What about the fossils and the dinosaurs?

For Neo-Darwinism to be realistic, fossils need to be millions of years old. Fossils are life forms stuck in the sediments. Sediments are laid down in strata and the only way that Neo-Darwinism can be true is that these sediments deposited gradually for millions of years that are recovering fossilized animals and plants that have been preserved until this day. Some species do not exist today and many are fascinated by the dinosaurs that supposedly existed millions of years ago. The deeper a life form is found in the strata, the older it is supposed to be. Methods for dating the age of fossils have been developed to evaluate the age of the strata and the species. The problem is that in the process of validating these dating methods, it was assumed that Neo-Darwinism was real. The fight between the Creationists and the Evolutionists has been so vitriolic that a very strong bias was installed to publish only the methods that fitted the evolution paradigm. All earth sciences were funneled into this paradigm. The result was a worldwide accumulation of support toward the Neo-Darwinian paradigm in terms of peer review articles. Those who dared oppose this paradigm were excluded from the white coats part of the Neo-Darwinism "religion." This movement was more based on an ideology than on a scientific method. Evolution became an obvious "fact" as opposed to a theory.

There is an alternative theory and the facts seem to fit with it. Briefly, in a message given to Rael[59] by the extraterrestrials, it is explained that after creating humans, a faction of the extraterrestrials educated our ancestors about

---

[59] www.rael.org

who we were and how we came into existence. They also educated them so much that a civilization capable of air and space travel developed (perhaps just in our solar system). But we were violent enough to scare the hell out of the extraterrestrial creators, who decided to stop this terraforming experiment and start over.

The faction that helped the newly born humanity to develop to this point copied the DNA of species that they wanted to save as well as groups of men and women to be reimplanted after the destruction. The destruction was colossal. Marcel Terrusse, a French chemical engineer, explained what this destruction would have looked like:

> When the CIA gave Hudson Institute the task of studying the world distribution of natural resources such as petrol, gas and coal, Professor Nehring who was put in charge discovered a rather strange phenomenon. He put together the continents as they would have been at the end of the geological tertiary period before they split up, and found that all the main oil springs such as those in the Artic and Alaska, the asphalt sands of Alberta, the bitumen schist of Colorado, Mexico and Venezuela, the heavy oils of "Orenoc", Nigeria, South Sahara, Libya, Arabia, Iran and Siberia all formed a circle.
>
> Present studies on petrol reveal that it is the result of what used to be living organisms, such as plants, animals etc., which have decomposed in anaerobic condition, that is to say, in the absence of air, where special bacteria reduce and transform their proteins
>
> Dead trees are the main source of these proteins and fats but what usually happens when one of them dies is that normal air breathing bacteria decompose it and recycle it into forest food chain without producing any petrol, so for the above to occur, the tree must be buried straight away to prevent access of air.

But when oil fields are examined, one finds that they are extremely deep (2,000 meters deep in northern France) and cover considerable areas (10,000 square miles in the Appalachians USA) so the actual volume of material, which must have been brutally buried all in one go, was enormous.

No theory up to now can satisfactorily account for these facts, but Raelians hold the key to this enigma.

When the extraterrestrial creators of mankind [called the Elohim in the Bible – a Hebrew word meaning "those who came from the sky" – mistranslated into God in English in the Bible; see www.rael.org] decided to destroy their bases, their laboratories and all that they had created on Earth, they must have had to use extremely powerful methods of destruction, which, as well as breaking up the original one continent and sending each respective fragment drifting outwards from the center of shock, must also have swept the whole land surface. Since the shock waves will have spread outwards from the bomb's point of impact, all the living matter including immense forests, animals and even man would have been blown away by this and buried deep, immediately and all together, under tons of earth in a ring shape, circling the central explosion. This would explain how extremely large volumes of living matter got buried quick enough to provide anaerobic conditions in the large circular formation that intrigued Nehring so much when he pieced back together again all the fragments of what was once our unique one continent.

During the time when the Elohim built the one continent until when they broke up, erosion would have accumulated sediment rich in animal matter such as corals and shells, into the oceans, especially around the edge of the continent. As, what later became the North and South America continental plates, scraped along the ocean bed, all the sediment built up to from the Andes and the Rocky Mountains range.

The same for the Indian sub-continent, which broke off Africa, and sliding northeast, trapped in its path enormous amounts of material, which we now know as the Himalayas. The Antarctic continent drifting south got covered with a huge coating of ice, preserving until today traces of tropical vegetation. Meanwhile Australia, part of Africa and India went southeast, accumulating along its edge sediment, which was to form the Australian Great Dividing Range. The destruction was cataclysmic and had provoked not only large geological changes, but climatic ones too, obliterating innumerable life forms, burying them under layers of sand, lime, earth and a sort of muddy ice, which preserved then until this day and from which they periodically emerged, released from their frozen tombs of the great Siberian North.

Only a small handful of humans were saved by the Elohim. Upon their return they found the continent completely unrecognizable and ravaged by the destructions, the geological upheavals having resulted in parts of the surface being decapitate and the appearance of cracks in the crusts layer causing volcanic activity.

It is easy to see how the story of the unique continent breaking up into many fragments "where there was once land but now is water" and vice versa got distorted into the legends of Atlantis or Mu where a continent is reputed to have disappeared into the sea. In fact, it did not sink, it merely moved aside.

Not all the living organisms were recreated after the flood; some considered to be monstrous and detrimental to the ecological balance such as the large reptiles and dinosaurs were left out, which explains the sudden and simultaneous disappearance of these anti-diluvian beasts.

After the "flood", the Elohim cohabitated with the humans of earth and the traces of their presence which

> we can still find scattered all over the world, are post diluvian.[60]

A couple of years after Marcel Terrusse's hypothesis was published (1978), Luis Alvarez and his son Walter, both geologists, suggested that a giant asteroid or comet hit the Earth approximately 65 million years ago and caused the mass extinction of the dinosaurs and over 70% of all life on Earth (numerous scientists are now supporting this view). "This bold proposition resulted from their discovery, near the medieval town of Gubbio, Italy, of a centimeter-thick clay layer. They found that the clay contained high concentrations of the element iridium, extremely rare in earth's crystal rocks but quite abundant in certain meteorites."[61] A decade later, a team of scientists discovered a 150-mile-wide crater near the village of Chicxulub on the Yucatán Peninsula, Mexico. A growing number of scientists believe the impact was responsible for the extinction of the dinosaurs and more than 70 percent of Earth's living species 65 million years ago.

Because of the naturalist bias scientists have, forcing any explanation via natural processes only, meaning without any intelligent intervention, the fact that we find high

---

[60] Intelligent Design – Message from the Designer, Raelian Foundation, 2005 (this excerpt was first published in the French original version "Accueillir les Extraterrestres" in 1978).

[61] Luis W. Alvarez, Walter Alvarez, Frank Asaro, Helen V, Michel, "*Extraterrestrial Cause for the Cretaceous-Tertiary Extinction,*" *Science*, 06 Jun 1980, Vol. 208, Issue 4448, pp. 1095–1108.

https://saturniancosmology.org/files/chicx/sharpton.txt

concentrations of iridium in craters formed by atomic explosions that the Alvarez team found in the clay never turned it into a causation candidate. Marcel Terrusse was the first to publish this theory (1978) and predicted implicitly the existence of craters related to mass extinction of life. These craters are inferred to be artificial because a bombing caused them, as revealed by the extraterrestrials in the message given to Rael. He also inferred that the continent broke up because of these impacts, and that all fossils, petrol fields and coal deposits are the results of this cataclysm.

According to the testimony based on the extraterrestrials' message, terraforming Earth occurred over a period of 13,000 years, 25,000 years ago, ending with the creation of seven races of human beings. Then it was destroyed in a few hours or days some few unknown number of thousand years after that. Layers of sediments are the results of the seawater being forced to go above the dry lands when the continent was breaking up and when the pieces were pushed outward by the explosion. The canyons were created when the water that mounted the continent pieces as they moved rushed out.

Petrol, coal and fossils are the results of living matter stuck in the earth after the cataclysm. The facts strongly indicate a cataclysmic scenario (dinosaurs biting each other and fossilized in their battle positions, mammoths with food still in their mouth, animals intertwined with trees, fossilized upside-down trees still with leaves on branches, etc., and this on all continents). This hypothesis makes sense given the actual evidence.

Militarism and UFOs

Some military secret ops are probably preparing to fight the extraterrestrials since their mentality leans in that direction. UFOs are automatically under the umbrella of national security since, for one, they are perceived as a possible threat or they are used that way to justify the secrecy. Also, a cover-up was deemed necessary for hiding the possible scientific technologies that could be "guessed" or reverse-engineered from contacts with ETs or from recovered UFO vehicles that seem to have crashed. The power in place is so imperialistic that dominating other nations is its primary goal. I would doubt that the possible UFO crashes would be accidental. I rather think that this would be part of tests to evaluate our reactions to their existence and their technology. These crashes should involve spaceships that possess just the minimum technology to make us realize that this is extraterrestrial technology. The extraterrestrials would then study what we do with it. Using extraterrestrial technology for wars would make us fail their evaluation. Who would be interested in contacting humanity in which countries are destroying each other for reasons of cupidity, for differences in skin color, for religions or for differences in societal organizations?

If ETs are looking to eventually share their knowledge with us, they will not be inclined to do so if we use technology to build weapons and kill people. I can imagine their fear of what we could do to them if we were to possess the same level of technology that they have. Indeed, being from another planet is a higher degree of difference than the difference between races and culture here on earth, bearing in mind that differences have been used to motivate wars on earth. I am convinced that the war mentality will make us destroy humanity if the good people of earth do not rise up, especially the top scientists.

How much would it take for the extraterrestrials to help us? Should we prepare for wars or be nice and help each other? Would you be more inclined to love and help cute kitties or enraged hyenas? It looks like peace on earth is our only logical and humane choice. In addition, perhaps being as cute and charming as we could be should help tremendously in that regard.

The extraterrestrials have created humans in their image and as beautiful as they can imagine, so I think we should use the most powerful cards we have in our deck: the charm and beauty of humanity on a path of genuine peace. We are their children, so we should expect them to love us more if we are beautiful in mind as well as in action.

The particular problem with humanity is that a section of humanity, which is very powerful, has been trying to get control of the whole world, leaving destruction and death in its wake. But the ET visitations have disrupted the plans of these villains. And since they will never give up their power, they will do everything in their power to control humanity even if it is to vilify ETs to the point of mobilizing the energy of humanity to wage war against our parents from space. Wernher von Braun had warned us about this exact strategy. The media is under the control of these men, who seek total control no matter how destructive their actions are. We need to get control of the media otherwise we will be brainwashed to behave in a way that will lead to our own destruction…

Conclusion

My own conclusion is that life in the universe is propagated as Enrico Fermi stated. Once a civilization reaches a level where it can travel into space and create life scientifically, the instinct pushes it to propagate. To give an analogy, if you give a brush and a canvas to a person that has just learned how to paint and enjoy doing it, that person will most likely create a painting at some point. Thus life spreads as an explosion that never ends, exponentially.

From the perspective that all the signs point to Earth having been terraformed by extraterrestrials it would be normal for these beings to contact us. People expect a spaceship to land near the White House, Moscow, Paris, etc., but surely the extraterrestrials have their own way to make contact. History teaches us that the method of choice for them is via contactees, or messengers. Religions, with their associated texts and prophets, attest to that method of contacting a civilization.

Among all the contactees, Rael is the only one who brings a complete message from extraterrestrials that fits the terraforming theory. This message also includes the keys to resolving our problem and surviving this forked road: that could lead to self-destruction or worldwide peace.[62]

This is why I have been helping Rael in his mission, which is to spread the message of those extraterrestrials who

---

[62] Intelligent Design– Message from the Designer, Raelian Foundation, 2005 (this excerpt was first published in the French original version "Accueillir les Extraterrestres" in 1978).

created scientifically all life on Earth. Rael created several nonprofit organizations that allow people to be part of this twofold mission. The Raelians are pure of heart and care for the future of humanity no matter the level of negativity within the mainstream media that has smeared Raelism since its inception.

Religions contain the traces of the contacts that humankind has had with these extraterrestrials, originally referred to as the Elohim ("those who came from the sky" or modernly "extraterrestrials") in the Hebrew Scriptures. Rael is considered by the Jewish members of the Raelian Movement as the Mashiach[63] as there is evidence for it. Rael, now living in Japan, has taken on the name *Maitreya Rael* since he is perceived by the Asian members of the Raelian Movement as the new Buddha as there is evidence for it.

Maitreya Rael is not a god. No one is perfect. He admits that he has made many mistakes, which is mostly a question of perception anyway. Most Raelians understand that making mistakes is part of growth. In many cases, making mistakes lead to great discoveries.

Besides his main mission, which is to spread the Elohim's message to humanity and to build an embassy for them — the organizations he founded have had great success for the first part — Rael has been focused on the journey to better himself while enjoying life as much as possible as well as to guide others along their own journey. Being Raelians is first understanding the message from those who created scientifically life on earth and then helping the messenger

---

[63] www.officialmessiah.org

who is the official representative of our parents from space. Caring about Maitreya Rael is caring about the Elohim who missioned him.

Rael and his followers have known the concept of "fake news" since 1974. I am happy that people are starting to wake up to this poison. The dogs may bark but the train keeps moving. No matter the smears, the Raelian Movement will continue to spread its message to humanity and, at some point, will build an embassy for our parents from space. And if our desire to meet them is strong enough, and if humanity is peaceful enough, they will land at the designated embassy. This will be the most powerful moment in the history of humanity…

In summary, one of the huge philosophical impacts of this new vision is to realize that there is no God and that life in the Universe is not a rare occurrence. The Universe should, in fact, be full of life according to the Fermi principle that we have shown in this writing. We are not alone, and our own existence is most probably linked to that of others out there in space who are watching us — our *parents* from space.

The most important thing for us now is to unite as one world and to eliminate wars in order to demonstrate to those who observe us that we have a sufficient level of wisdom among us to not be considered a threat to others in the cosmos. Our survival depends on it on multiple levels. The way we behave among nations is a predictor of how we will behave toward extraterrestrials and they know it.

For that we need to elect the best among us, not the most "stupid" or the most "normal".

So let us be very aware of this when devising behavioral strategies in anticipation of extraterrestrial visitations. We should not base these strategies on the obtuse views that incorrectly assume that ETs are as primitive as we were as a human society at the time of the "discovery" of the Americas by Europeans, which was not a good thing for the native people. This is completely different because we are the extraterrestrials' children.

I can even see our parents from space protecting us from dangers that we cannot control (solar flares, big meteors, worldwide volcanic eruptions, etc.). That would better explain why big meteors explode before hitting ground (sometimes seen accompanied by a UFO before the meteors explode).

On earth, except for some violent secret "factions" that control many governments, our diverse societies are gradually evolving towards peace and this appears to be a natural tendency of human existence. While much more needs to be done, we can predict that we will become even more respectful of our differences and continue to progress toward world unification over time. If we do not do this, we will self-destruct and eliminate ourselves from the universe. Therefore, because of this principle, we should expect that an extraterrestrial civilization that is advanced enough to travel to our planet would also be peaceful as it would have successfully passed the test of self-destruction.

Our efforts to establish peace on earth, which is an indication of our "coming of age," would not go unnoticed by those who are observing and monitoring our progress from above. Our civilization is likely being evaluated on this very challenge: our capability to control our own violence and aggression toward others. This capability is directly linked to our capacity to live together in peace and harmony as a human society. If we can accomplish this ideal situation, it will increase the likelihood that we would have an equal capacity to show genuine respect and love to whoever comes to visit us from other worlds. This would in turn increase the level of trust that an extraterrestrial civilization could invest in us. If we help ourselves toward refinement and harmony, chances are that those who are observing us from space will be good to us. Without love, there is no future for us as a human society from whatever angle we look at it.

> How would we explain the global arms race to a dispassionate extraterrestrial observer? How would we justify the most recent destabilizing developments of killer-satellites, particle beam weapons, lasers, neutron bombs, cruise missiles and the proposed conversion of area the size of modest countries to the enterprise of hiding each intercontinental ballistic missile among hundreds of decoys? Would we argue that ten thousand targeted nuclear warheads are likely to enhance the prospect for our survival? What account would we give of our stewardship of the planet Earth? We have heard the rationales offered by the superpowers. We know who speaks for the nations. But who speaks for the human species? Who speaks for Earth?[64]

There are important changes to be made before it is too late. Let us use science spiritually, and for peace and love only. In times of crisis we use the best amongst us. This time is now. The best of the best minds amongst us,

---

[64] Loc. cit.: Sagan p. 329.

especially among the scientists, should rise to this occasion and to assemble to save humanity.

In the final analysis, we are one People on Earth! Uniting as one society should be a priority perhaps in a planet federation of countries. There is nothing more urgent. Unfortunately, many scientists are like this kid who plays video games in the basement while the house is on fire. There are smoke and screams but the kid wants to finish the game!!! It is almost midnight on humanity's doomsday clock for there are only a few minutes or even a few seconds left before the clock sounds humanity's doom.

Scientists, could finally the best of you leave the joystick and help save humanity? It should be your first mission.

Index

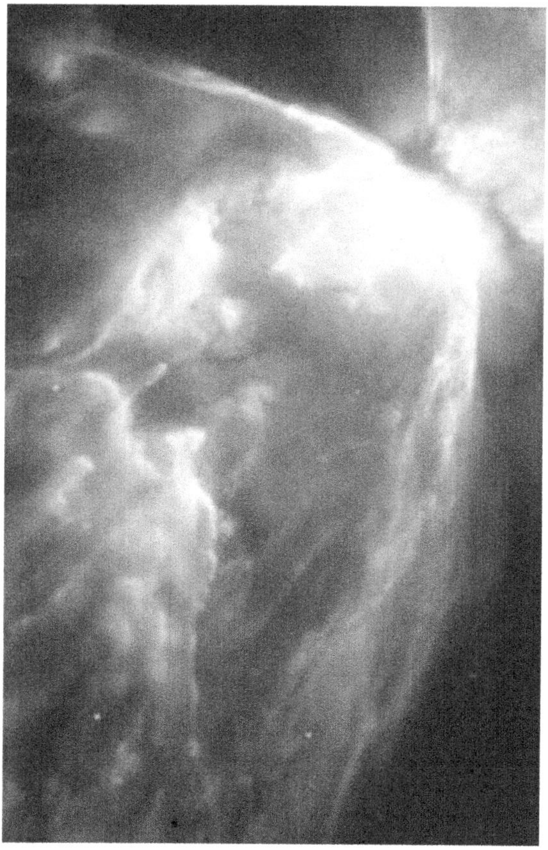

The Bug Nebula (NASA)

Made in the USA
Monee, IL
16 June 2020

33316372R00052